FUSION

FARMER LLAMA'S FARM MACHINES
TRACTORS
BY KIRSTY HOLMES

BEARPORT
PUBLISHING

Minneapolis, Minnesota

Library of Congress Cataloging-in-Publication Data is available at www.loc.gov or upon request from the publisher.

ISBN: 978-1-64747-547-5 (hardcover)
ISBN: 978-1-64747-554-3 (paperback)
ISBN: 978-1-64747-561-1 (ebook)

© 2021 Booklife Publishing
This edition is published by arrangement with Booklife Publishing.

North American adaptations © 2021 Bearport Publishing Company. All rights reserved. No part of this publication may be reproduced in whole or in part, stored in any retrieval system, or transmitted in any form or by any means, electronic, mechanical, photocopying, recording, or otherwise, without written permission from the publisher.

For more information, write to Bearport Publishing, 5357 Penn Avenue South, Minneapolis, MN 55419. Printed in the United States of America.

IMAGE CREDITS

All images are courtesy of Shutterstock.com, unless otherwise specified. With thanks to Getty Images, Thinkstock Photo, and iStockphoto. Cover - NotionPic, Tartila, A-R-T, logika600, BiterBig, Hennadii. Aggie - NotionPic, Tartila. Grid - BiterBig. Farm - Faber14. Spreaders - Hennadii H. 2 - alazur. 4 - Hennadii. 5 - Mascha Tace. 6 - alazur. 7 - alazur, stockakia, vulcano. 8 - Faber14. 9 - Faber14, Mayabuns. 10 - barkarola, Murad_Mammadov, JustPixs. 11 - Janis Abolins, Panda Vector. 12&13 - AndriyA. 13 - Ira Yapanda, Marina Akinina, K-Nick, yafi4. 14&15 - Paul Kovaloff. 16 - vectorisland, momoforsale, Africa Studio. 17 - KVArt, piscari. 18&19 - Alfazet Chronicles, alazur. 20 - Mascha Tace. 21 - DRogatnev. 22 - Janis Abolins, RedKoala, Motorama, GordanD. 23 - alazur, Ralf Geithe.

CONTENTS

Down on the Farm! 4

What Is a Tractor? 6

Before the Tractor 8

An Iron Horse 10

Parts of a Tractor 12

A Tool for a Tractor 14

Driving a Tractor 16

Record Breakers 18

Get Your Llama-Diploma 20

A Nice Glass of Llamanade 22

Glossary 24

Index 24

DOWN ON THE FARM!

Welcome to Happy Valley Farm. You must be the new **farmhand**. I'm Aggie, and I'm a farmer llama.

She's an expert in her field!

Before you start work in the fields, I'll make sure you know all about the different machines here on the farm. Let's get you trained in!

What You Need to Know

- [ ] Where the driver sits!
- [ ] All the jobs a tractor can do!
- [ ] How to steer!
- [ ] What an iron horse is!

WHAT IS A TRACTOR?

Time to check on the lemon trees! Happy Valley Farm is known for its llamanade, you know!

Tractors are working machines on a farm. They are built to be strong and sturdy.

6

Tractors do lots of jobs around the farm, and they are good at pulling heavy **loads**. They are very important machines for every farmer.

BEFORE THE TRACTOR

Before tractors, all farm work had to be done by humans and animals. Large, strong animals were used to **plow** fields and pull things. These animals were called draft animals.

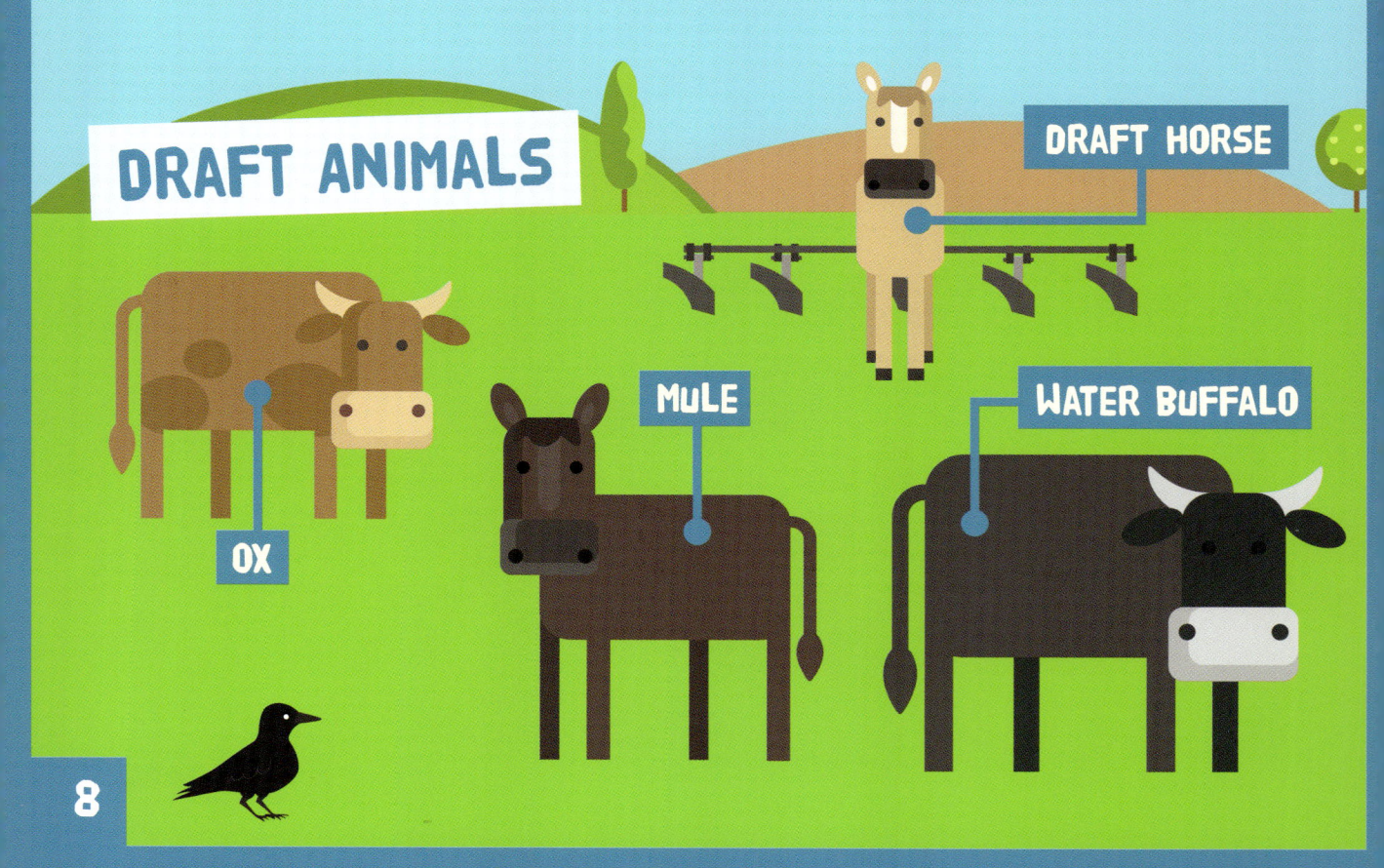

DRAFT ANIMALS

- DRAFT HORSE
- OX
- MULE
- WATER BUFFALO

For really tough jobs, a team of draft animals could be used. Two or more horses or oxen would have been **hitched** together to pull a plow through the fields.

FARMER

OX

PLOW

MULE

The more animals you have on the team, the more they can pull!

AN IRON HORSE

During World War I (1914–1918), many horses were needed for the war. But farmers still needed help on their farms. Tractors, sometimes called iron horses, soon became popular.

We measure the power of tractors in horsepower. It tells us how much work the tractor can do based on how many horses it would take to do the same job.

Modern tractors can have up to 620 horsepower! That's a lot of horses...

PARTS OF A TRACTOR

Let's look at the parts of a tractor.

SEAT
This is for the driver.

EXHAUST
The waste from the engine puffs out here.

WHEELS
The deep **tread** on the tires helps tractors grip fields.

12

CAB
A cab with walls protects the driver.

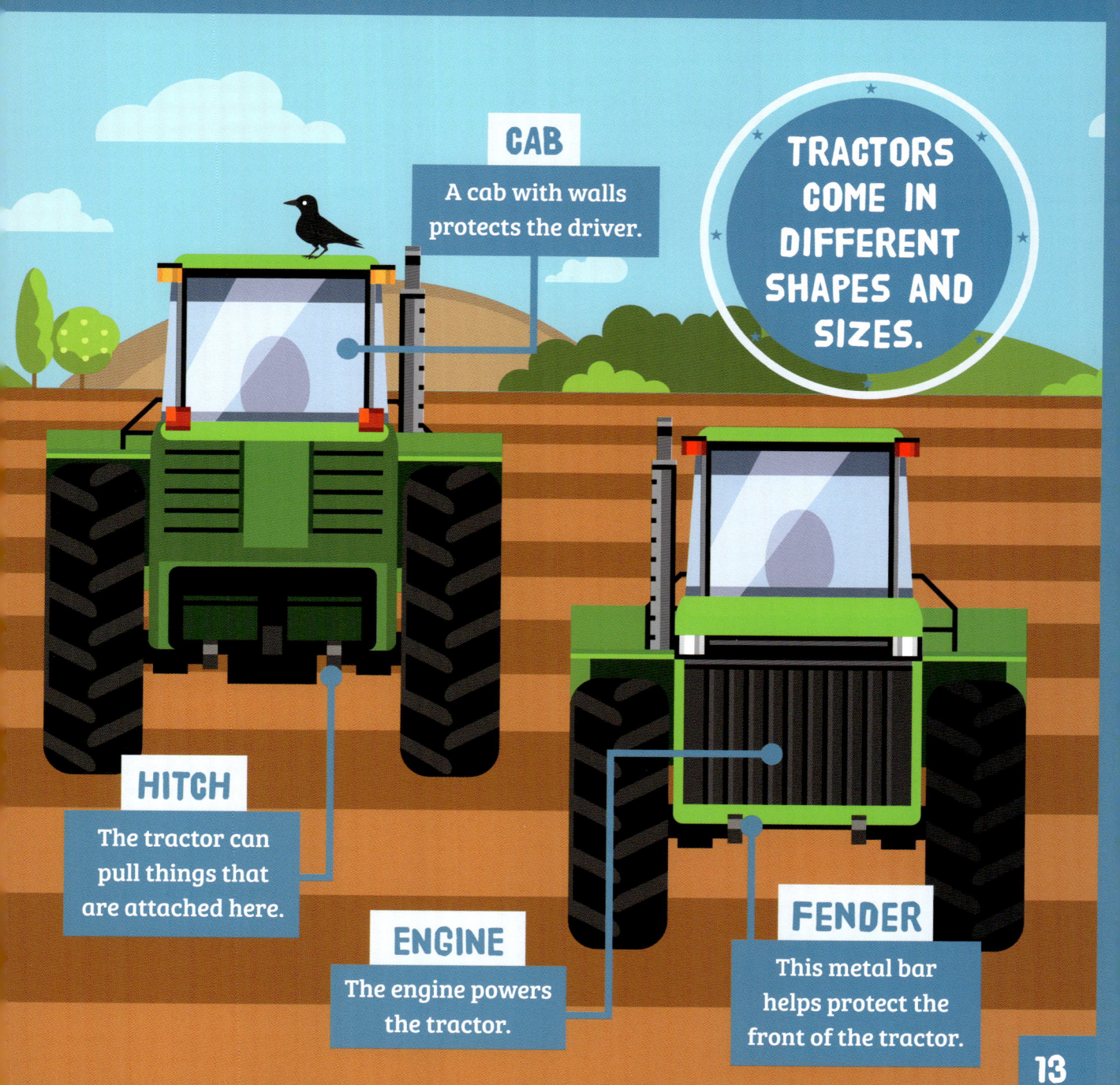

TRACTORS COME IN DIFFERENT SHAPES AND SIZES.

HITCH
The tractor can pull things that are attached here.

ENGINE
The engine powers the tractor.

FENDER
This metal bar helps protect the front of the tractor.

13

A TOOL FOR A TRACTOR

Different tools can be attached to the hitch, and then the tractor can do different jobs around the farm.

HARROW
Harrows make the soil smooth.

PLOWS
Plows dig deep into the soil to make it ready for the farmer to plant seeds.

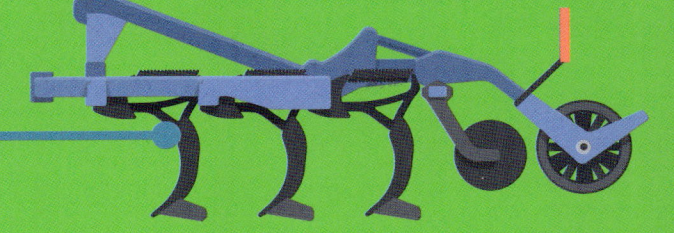

CULTIVATOR
Cultivators turn the **topsoil**. This makes the soil better for new seeds.

14

"This is a great lemon **harvest**! We can make so much llamanade!"

Tractors aren't very fast, but they are very strong! They can be hitched to a trailer to pull heavy loads.

DRIVING A TRACTOR

Different tractors have different controls. Let's see some of the basics that most tractors have . . .

The farmer can drive the tractor with the steering wheel.

The brake slows the tractor down.

The main switch turns the tractor on.

The gear lever controls how much power goes to the wheels.

The throttle speeds the tractor up.

16

Tractors of the future might not need a driver at all. The farmer would set a computer to tell the tractor what to do, and then off it would go!

Now, I have more time to do other jobs!

RECORD BREAKERS

On May 8, 2016, Hubert Berger set off from Germany in his blue tractor with a little gray camper hitched on the back. Hubert had quite a trip ahead of him.

Hubert's tractor was built in 1970!

In six months, Hubert and his tractor visited 36 countries all around Europe. Hubert's trip broke the world record for the longest journey by tractor.

Hubert traveled 15,700 miles (25,300 km).

GET YOUR LLAMA-DIPLOMA

Well done, farmhand! You made it through the training. If you've been paying attention, this test should be no prob-llama...

Questions

1. What are animals that work on farms called?

2. How do we measure the power of a tractor?

3. What is special about a tractor's wheels?

4. What does a harrow do?

5. Where does the driver sit in a tractor?

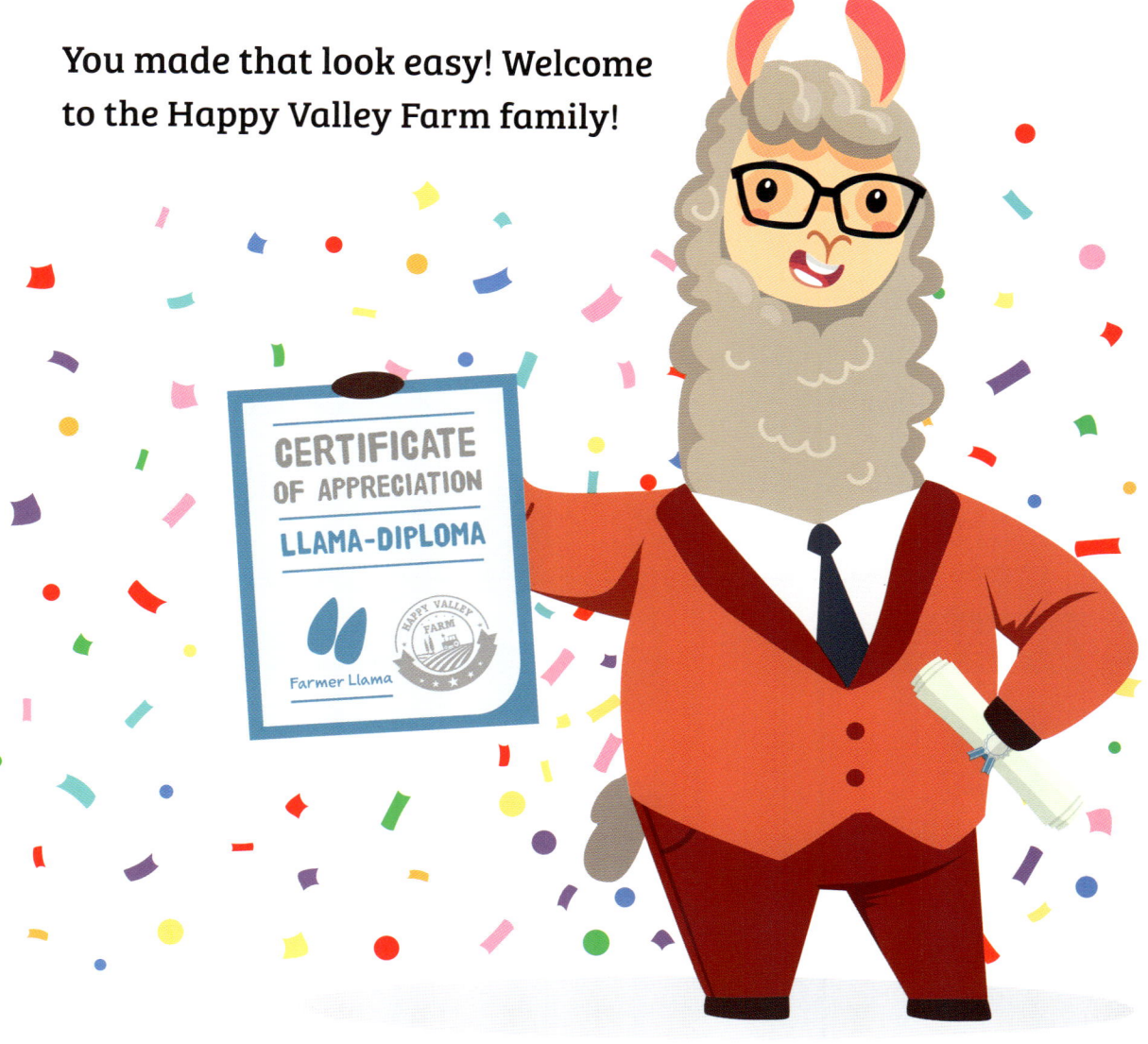

You made that look easy! Welcome to the Happy Valley Farm family!

Download your llama-diploma!

1. Go to **www.factsurfer.com**
2. Enter "**Tractors**" into the search box.
3. Click on the cover of this book to see the available download.

Answers: 1. Draft animals 2. Horsepower 3. They help the tractor grip the field. 4. Smooths the soil 5. The cab

21

A NICE GLASS OF LLAMANADE

At the end of a hard day, we like to have some fun. Even with the tractors helping us out, farming is thirsty work!

STEP ONE
Find a tractor

STEP TWO
Fuel up with llamanade

STEP THREE
Grab your helmet

GLOSSARY

FARMHAND a person who works on a farm

HARVEST the process of gathering crops

HITCHED joined or attached to something

LOADS amounts of things carried or pulled

PLOW to turn over the soil to make it ready to plant seeds

TOPSOIL the upper layer of soil that has the most nutrients

TREAD the grooves on a tire that help it to grip

INDEX

BERGER, HUBERT 18–19
CULTIVATORS 14
DRAFT ANIMALS 8–9
HARROWS 14
HORSEPOWER 11
PLOWS 8–9, 14
TRAILERS 15
WORLD WAR I 10